PowerKiDS Readers
THE UNIVERSE

THE MOON

Elisa Peters

PowerKiDS press
New York

Published in 2013 by The Rosen Publishing Group, Inc.
29 East 21st Street, New York, NY 10010

First Edition

Editor: Amelie von Zumbusch
Book Design: Kate Laczynski

Photo Credits: Cover, pp. 4, 6, 8, 12, 14, 16, 18, 24 (earth, moon) Shutterstock.com; p. 10, 24 (crater) Richard Wahlstrom/Getty Images; p. 20 O. Louis Mazzatenta/National Geographic Image Collection/Getty Images; p. 22 NASA NASA/Getty Images.

Library of Congress Cataloging-in-Publication Data

Peters, Elisa.
The moon / by Elisa Peters. — 1st ed.
p. cm. — (Powerkids Readers: The universe)
Includes index.
ISBN 978-1-4488-7386-9 (library binding) — ISBN 978-1-4488-7466-8 (pbk.) —
ISBN 978-1-4488-7538-2 (6-pack)
1. Moon–Juvenile literature. I. Title.
QB582.P48 2013
523.3—dc23

2011044833

Manufactured in the United States of America

CPSIA Compliance Information: Batch #CS12PK: For Further Information contact Rosen Publishing, New York, New York at 1-800-237-9932

CONTENTS

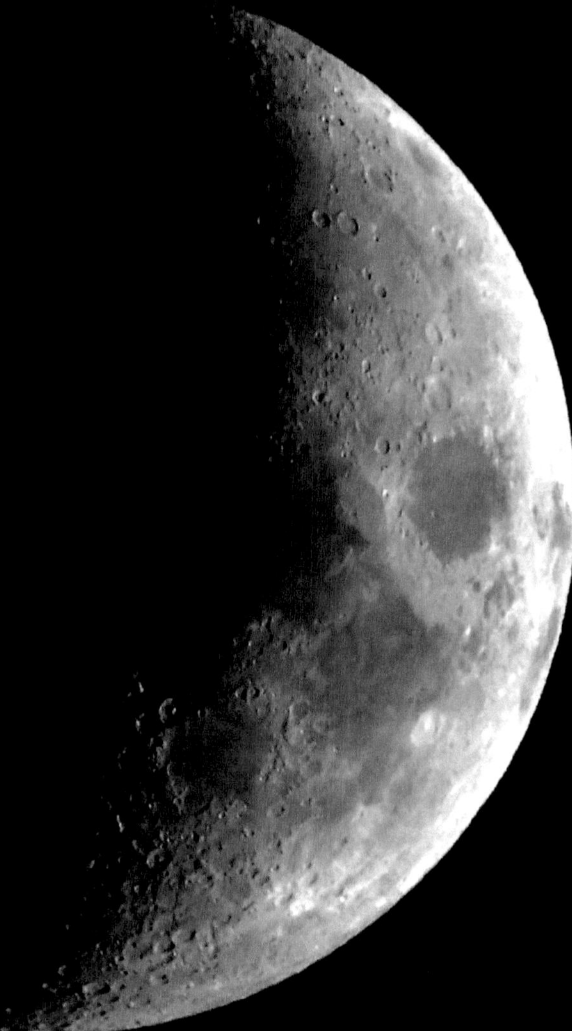

4

The **Moon** is made of rock.

It circles **Earth**.

It is smaller than Earth.

It has holes called **craters**.

There is no air there.

The Moon has phases.

Some nights it is full.

Some nights you see only part of it.

People went to the Moon.

They walked on it!

WORDS TO KNOW

crater

Earth

Moon

INDEX

WEBSITES

Due to the changing nature of Internet links, PowerKids Press has developed an online list of websites related to the subject of this book. This site is updated regularly. Please use this link to access the list:
www.powerkidslinks.com/pkrtu/moon/